Lincolnshire
COUNTY COUNCIL

discover libraries

**This book should be returned on or before
the last date shown below.**

To renew or order library books please telephone 01522 782010
or visit www.lincolnshire.gov.uk
You will require a Personal Identification Number.
Ask any member of staff for this.

A HANDBOOK

TO THE

MARINE AQUARIUM:

CONTAINING

PRACTICAL INSTRUCTIONS FOR
CONSTRUCTING, STOCKING, AND MAINTAINING A TANK,
AND FOR COLLECTING PLANTS AND ANIMALS.

BY

PHILIP HENRY GOSSE, F.R.S.

Second Edition, Revised.

LONDON:
JOHN VAN VOORST, PATERNOSTER ROW.
M DCCC LVI.

Gosse, Philip Henry (1810-1888)
A Handbook to the Marine Aquarium:
containing practical instructions for constructing, stocking, and
maintaining a tank, and for collecting plants and animals
1856, second edition revised

This facsimile edition published in 2010
by Euston Grove Press
20 Elderwood Place, London SE27 0HL United Kingdom
www.EustonGrove.com

British Library Cataloguing in Publication Data

Gosse, Philip Henry, 1810-1888.
  Handbook to the marine aquarium.
  1. Marine aquariums.
  I. Title
  639.3'42-dc22

ISBN-13: 9781906267186

LONDON:

R. CLAY, PRINTER, BREAD STREET HILL.

PREFACE.

THE increasing popularity of the Marine Aquarium demands a Handbook of Practical Instructions for establishing and maintaining it, and I am probably the most proper person to write it. Perhaps it might have been sufficient to refer inquirers to my volume on the subject; but the price of that work, arising mainly from the costliness of its illustrations, puts it beyond the power of many persons, who yet desire to keep marine animals. The main portion of that volume is, moreover, occupied with the habits and manners of the tenants in an Aquarium.

The concluding chapter of that work has formed the ground of the present Handbook. The whole, however, has been re-written, and copious additions have been made, bringing it up to the present state of our experience. The price at which it is issued will, it is hoped, bring it within the reach of all.

P. H. G.

LONDON,
September, 1855.

PREFACE

TO THE SECOND EDITION.

THIS Edition has been carefully revised; a few paragraphs, that seemed needless, have been omitted, and others of new interest have been added. P. H. G.

July, 1856.

CONTENTS.

HANDBOOK

TO

THE MARINE AQUARIUM.

An Aquarium is simply a vessel of water in which aquatic plants, or animals, or both, are preserved alive; and a Marine Aquarium is, of course, such a vessel, whose contents, animate and inanimate, are derived from the sea.

In general it is the marine Animals that form the main source of interest, everything else being merely accessory to these. Many of the sea-plants, "weeds" though they are called, are indeed very beautiful; the elegant forms of some, the delicate muslin-like tracery of others, the plumose lightness of more, "fine as silkworm's thread," and the beautiful play of colours, red and green, which a well-stocked Aquarium displays, as the light is transmitted through their pellucid substance, may claim for these objects more than an indirect attention. Still it is true, that, in most cases, they are preserved because they cannot be dispensed with.

If we attempt to collect and to keep marine

animals alone in sea-water, however pure it may have been at first, it speedily becomes offensively fetid, the creatures look sickly, and rapidly die off, and we are glad to throw away the whole mass of corruption.

Why is this? Why should they die in our vessels, when they live so healthily in the little pools and basins of the rock, that are no larger? For the very same reason that we should quickly die in a room perfectly air-tight. The blood of all animals requires to be perpetually renewed by the addition to it of the element called oxygen; and when it cannot obtain this it becomes unfit for the support of life. Terrestrial animals obtain this gaseous element from the air; aquatic animals (that is, those which are strictly such) obtain it from the water. But in either case it is principally produced by *living plants while under the action of light.* If, then, we can furnish our captives with a perpetual manufactory of oxygen, the main cause of their sudden death is removed. Of course they have other requirements, but this is the most urgent, the indispensable.

In a state of nature, the rocks, the crannies, the pools, the sea-bottom are studded with various living plants, which we call sea-weeds; and these, under the daily stimulus of sunlight, direct or indirect, produce and throw off a vast quantity of oxygen, which, by the action of the waves and currents, is diffused through all parts of the habitable sea, and maintains the health of its countless swarms of animals.

In an Aquarium we seek to imitate this chemistry of nature. We collect the plants as well as the animals; and, a little observation teaching us how to proportion the one to the other, we succeed in maintaining, on a small scale, the balance of animal and vegetable life. Other less important benefits result from this arrangement; the creatures love retirement and shelter, and this they find in the umbrageous fronds; they delight to roam and to play and to rest in the feathery tufts, and not a few find their favourite food in the delicate leaves of the submerged herbs.

On the other hand, the plant is indebted to the animal for some of its supplies. The carbon, with which its solid parts are built up, is derived from the carbonic acid which is thrown off by animals in the process of breathing;—a poisonous gas which would soon vitiate the water were it not taken up and appropriated by the plants.

Such, then, is the principle on which the Aquarium is founded; and any conditions under which it can be carried out will serve, provided of course they be suitable in other respects to the habits of the animals and our purpose in keeping them. I now proceed to other details on the practical management, so far as I am able, from my own experience, and that of not a few of my friends, to give them; premising that I have at present (April, 1856), at my residence at Islington, one marine tank full of animals and plants in the highest condition, the water in which, though as clear as crystal and quite colourless, has never been even removed from the vessel since it was first put in, two years and two months ago. I have, also, other tanks and vases, which are from two years to one year old. The successful establishment of these has not been achieved without some failures and losses, which yet must not be considered as unmitigated misfortunes, since they have added to my experience, and better fitted me to understand and sympathise with the difficulties of other beginners.

THE TANK.

FORM, SIZE, AND MATERIALS.—So much depends on individual taste and means in this respect, as well as on the situation which the Aquarium is intended to occupy, that no rule can be laid down for dimensions. My largest tank (now in use as a *fresh* water Aquarium,) is a parallel-sided vessel, two feet long, one and a half foot wide, one and a half foot deep; the sides and ends of plate glass, three-sixteenths thick; the bottom, a slab of slate one inch thick; the corners of birch-wood, turned into pillars, each surmounted by a knob, and united by a frame-top or bar going all round. The glass is set in grooves in the slate and wood, and fastened with white-lead putty. I have two others agreeing with this in all respects, except in dimensions, the smallest being (in the clear) fifteen inches long, twelve inches wide, and twelve inches deep. This is a very neat and pretty object for a parlour table, and will hold at least fifty animals appreciable to the senses, provided they be well selected, and a sufficient growth of plants established.

I have also another tank with a sloping back, made on Mr. Warington's plan. It is of zinc, with the back and two sides of slate, the front and two front-sides alone being of glass. Its form is six-sided, the front resembling a bow-window; it is three feet long by one foot ten inches in greatest width, and the same in depth; the peculiarity is that the back slopes inward, so that the bottom is but eight inches wide. I cannot commend this form; its appearance is lumbering and inelegant; the opacity of the back and sides throws the interior into a degree of darkness, (even though placed in a south window,) which greatly impedes observation; and I cannot see, by comparison with my others, which are of glass all round, that the advantage anticipated, of admitting the light only from above, is real, or at least of sufficient importance to compensate the uninviting gloominess alluded to. Its depth also is too great; a foot of water is as much as is needful for a parlour Aquarium.

A novel mode of constructing tanks has been suggested to me by Mr. W. Dodgson of Wigton, Cumberland, which, as I have not tried it, I will describe in his own words.

"I have lately been constructing two Aquariums, and as the way in which they are made may be new and perhaps useful to you, I take the liberty of communicating it. Slate we have no opportunity of procuring in suitable pieces for joining, and our stone is too porous for the purpose. I therefore got the bottom and two ends made in one piece out of the yellow clay used for garden vases, chimney-tops, and other coarse pottery, and found it answered exceedingly well and has several advantages.

"Partly as a means of supporting the ends, but principally to form artificial rockwork and shelter for the animals, the two ends are buttressed inside with very rough pieces of clay, put on so as to leave plenty of holes and fissures; the fire fastens these firmly together and makes them as hard as stone. Grooves are left along the bottom, and up the two ends, for the glass to fit into. The outside is relieved with ornamental work. Cheapness is a great recommendation; the pot being charged 1*d*. or 2*d*. per hundred cubical inches, according to the amount of ornament. I should think your London potters would make something very neat, and the mould once made, they could be supplied very cheaply; and considering their freedom from leakage and their strength, I think they would prove as satisfactory as any kind of cistern. Mine are about three feet long, and thirty inches high and broad, holding thirty gallons each. I bed the glass with white lead, leaving about a quarter inch in depth of the groove unfilled. When the putty is set, I fill it up with shell-lac dissolved in naphtha and made into a paste with whiting. This sets very quickly, becoming as hard as stone; it is quite insoluble, and prevents the water from coming in contact with the lead, which I think must constantly be giving off small quantities of oxide."

In reply to some inquiries of mine, the same gentleman writes me further as follows:—

"I have delayed replying to you respecting the price of pottery for Aquariums till I had an opportunity of seeing the party who made mine. The price of one such as you name would be 6*s*. or 8*s*., but, as it would weigh four hundredweight, the carriage to London would be quite as much. He was rather doubtful whether with his coarse clay he could make one the length you name to hold together; as two or three of mine, which were shorter, separated in the middle when being burnt, from the great weight of the ends preventing the centre contracting regularly. He thinks your best plan would be to get one of the terra-cotta workers to make one, which he thinks he would do more cheaply than it could be sent from here; and their clay being finer, a much neater article could be made. It would be well to be on the spot, and see the rough clay put on the ends, as workmen in general have not much idea of what is required.

"To prevent the water filtering through the pots, mine were well glazed inside with flint-glaze, free from lead. Probably it would be better to glaze the outside, and leave the inside porous."

If a cylindrical form of vessel be preferred, it can be obtained without any material but glass in the construction. A very pleasing Aquarium, which has the advantage of cheapness, is greatly in request, formed of a propagating glass turned upside-down, and set on a stand of polished mahogany or rose-wood. This form has cylindrical sides, and a rounded bottom, terminating in a knob, which being inserted into a cavity in the stand, gives security to the whole. The bottom should be covered with sand or earth to a height sufficient to bring it up to the level of the cylindrical sides, for the convenience of observation. Vessels of this shape are now made up to 20 inches in diameter.

Confectioners' show-glasses are very suitable for small specimens; they are usually about twice as high as broad, and are therefore the more graceful. These afford peculiar facilities for the examination of their contents with a lens, as they can be easily moved round, and may be placed before a candle or lamp for nocturnal investigations. They may be had of various sizes, from three to eight inches

in diameter; six inches is a very convenient size.
I have made very pretty vases for minute objects
by taking the glass chimney of an Argand lamp,
and fastening a well-fitted cork into one end for a
bottom, on which I then poured black sealing-wax
varnish until a smooth water-tight surface was
formed. These are more convenient than wide-
mouthed phials, as the sides are more truly
perpendicular, and we avoid the unequal refraction produced by the thickened bottom of a
phial.

For a conservatory, to which an Aquarium would
form an appropriate accessory, a vase-like form
might be given to a large tank. If the outline
were octagonal, the objects in the interior would be
visible through the plate-glass, without the dis-
tortion caused by unequal refraction, which is a
great objection to vessels with curvilinear sides.
But in such a situation, the chief point of view
would be from above the surface of the water;
hence the depth should be comparatively small,
and the sides might recede, so as to increase the
width upwards.

A good many animals, especially Anemones,
Madrepores, Crabs and Annelides, may be kept in
broad and shallow pans, in which the water does
not exceed three or four inches in depth. The
glass pans used for milk are good for this purpose.
I have an earthenware pan with upright sides,
about five inches deep, in which I have imitated
the broken interior of a rock-pool, with irregular
projections and promontories of cement. One advantage of such a vessel is that sea-weeds may
almost be dispensed with, the extensive surface of
water absorbing a large quantity of oxygen from

the air. An oval foot-bath of whiteware makes a capital Aquarium.

COVERING.—Within an inhabited room, or wherever there is much liability to dust or soot, as there is necessarily everywhere in cities and large towns, the Aquarium may be protected by a cover. This may be made of fine muslin, or, which is better, of plate or sheet-glass, according to the dimensions required. The latter may simply be laid over the top of the vessel, allowing the escape of gases under the edge. It should be occasionally lifted for a moment, to allow of a change of the superincumbent air:—the necessity of this will be manifest, from the close smell which is perceived on lifting the cover, especially if there be many sea-weeds in the tank.

In ordinary circumstances, however, there is no necessity for a covering of any kind. My own tanks, though placed in an inhabited room, remain for months together uncovered, in winter and summer, without the least loss of transparency. The dust speedily sinks, and is harmless.

ASPECT.—The free access of light to the plants is indispensable; and therefore that situation is the best where the sun's rays fall most freely on their leaves. It is beautiful to see the thousands of tiny globules forming on every plant, and even all over the stones, where the infant vegetation is beginning to grow; to see these globules presently rising in rapid succession to the surface all over the vessel, and to see this process going on uninterruptedly as long as the rays of the sun are uninterrupted.

Now these globules consist of *pure oxygen*, eliminated by the vegetation under the stimulus of light; and as this is the vivifying principle of animal life, the importance of the process will be readily acknowledged. The difference between the profusion of oxygen-bubbles produced on a sunny day, and the paucity of those seen in a dark, cloudy day, or in a northern aspect, is very marked.

HEAT.—Yet there is one caution required. In summer the heat of the solar rays is very great, as well as their light; and if the vessel be small, and the volume of water very limited, it will become tepid in the mid-day sun, and the animals will be killed. Hence in a fierce summer day, it will be desirable to draw down the window-blind, or to interpose a curtain of muslin, oiled-paper, or ground glass, which will break the full power of the rays, without greatly interfering with their illumination.

On this subject a suggestion made by Mr. George Guyon in the "Zoologist" for March, 1856, is worthy of attention. "Since photography has become a popular science, it is pretty generally known that the three principles existing in common light,—luminosity, heat, and chemical action,—are to a great extent separable, and reside respectively in the yellow, red, and blue rays of the spectrum. It is moreover, I believe, considered that growing plants decompose carbonic-acid, and liberate the oxygen under the influence of the luminous, or yellow rays: if this latter opinion is correct, would not the interposition of a screen of *yellow glass*, while giving free admittance to the purifying influence, effectually prevent the water from getting over heated, by arresting the progress of the red or heat giving rays?"

THE PREPARATION.

Your Aquarium being brought home, fixed in its intended situation, and properly seasoned, the next thing is to fit it up as a dwelling for its living inhabitants. Two or three points may be noticed here.

ARTIFICIAL ROCKS, CORALS, &c.—When the two longer sides only of the Tank are of glass, the two ends being made of slate, the latter should be veiled, by being made to imitate the irregular projections and ledges of rock, which may be done in a very picturesque manner. For this purpose, Roman, Portland, or other cement which hardens under water, should be employed; the slate must be faced with this, which while plastic may be fashioned into the semblance of rock. Pieces of branching corals may be set in it, if the effect of such accessories be thought desirable, and cavities may be formed here and there, into which the fragments of stone that support growing sea-weeds my afterwards be dropped, so that the tufts may droop elegantly from the mimic cliff. A more elegant way of appropriating branching corals, is to make a broad foot of cement to them, plunging the base of the branch in it while soft; these, when the cement has hardened, will stand on the floor of the tank like trees, and imitate more perfectly the mode of growth of the arborescent madrepores.

Whenever cement is used, it will be absolutely necessary to allow it to remain in water for at least a month, in order to soak out the free lime, before it be introduced into the water which contains

animals. The water in which it is soaked should
be frequently changed, and as long as any pris-
matic scum appears on the surface, the cement is
unfit for use. I have known a whole consignment
of animals killed in one night from a neglect of
this precaution.

THE BOTTOM.—As very many marine animals
burrow, and as the observation of their proceedings
is very interesting, they should be provided with
the means of gratifying their inclinations. For
this purpose a layer of sand should be put on the
bottom of the tank, which may vary in depth from
one to three inches. If sand from a sea-beach can
be readily obtained, it is the most suitable; but
the next best is coarse river-sand, such as the
Thames sand commonly sold at the stone-wharves
of London for building purposes. It should be
well washed, until the water runs away clean:
fresh water will do very well for this, but it must
be drained off before the sand is put in. What is
called silver-sand, and the common yellow earthy
sand sold in the shops for scouring, are not at all
suitable, as they will tinge the water after any
amount of washing, the former with lime, the latter
with ochre.

Small pebbles or fine gravel, likewise well
washed, may be used to vary the bottom with the
sand.

Masses of rock, of dimensions suitable to the
Aquarium, should be put in to afford shelter and
concealment to such animals as like the gloom.
To afford this in the highest degree, a flat piece
may be set, like a table, or *cromlech*, upon two or
three upright blocks; or two tall pieces may lean
against each other, forming a rude arch;—care

being taken, whatever arrangement be chosen, that the masses stand with stability. It is of little consequence what sort of rock is selected,—limestone, sandstone, granite, conglomerate,—but the rougher, and the more full of cavities and angles, the blocks are, the better will be the effect.

WATER.—The purity of the water is of great importance. In London, sea-water may be easily obtained, by giving a trifling fee to the master or steward of any of the steamers that ply beyond the mouth of the Thames, charging him to dip it in the clear open sea, beyond the reach of rivers. I have been in the habit of having a twenty gallon cask filled for me, for which I give a couple of shillings.

The vessel in which it is conveyed requires attention. A cask is the best, if a considerable quantity of water is required; but it is absolutely indispensable either that it should be new, or at least that nothing injurious should have been previously contained in it, such as spirits, wine, chemicals, acids, &c.; since no soaking will prevent hurtful qualities from being communicated to the water. Even the bungs ought to be new. I knew an instance in which a consignment of animals was lost, from no traceable cause, except that the water-cask, which was quite new, had been stopped with a bung, which had been previously used in a jar of some chemical solution; yet the bung had been, as was supposed, *thoroughly* soaked and cleansed. If a cask of fir-wood can be procured, it is preferable: the wood of the oak, of which wine-casks are usually made, gives out *tannin* or *gallic acid*, to the contained water, which by its astringency converts the animal integuments into leather.

If the water on coming out of the cask has a brown tinge, without interfering with its transparency, this is suspicious. If you cannot get any other than an oak cask, let it be well seasoned for two or three weeks before it be used, by filling it with water (fresh or salt), changed every day.

For smaller quantities of water large jars of stone-ware are the best, being free from every objection arising from liability to taint or tinge. Both casks and jars can be easily sent by railway to any part of the kingdom; and pure water will not spoil by delay.

ARTIFICIAL SEA-WATER.—In July, 1854, I published the following communication in the "Annals and Magazine of Natural History."

On Manufactured Sea Water for the Aquarium: — " The inconvenience, delay and expense attendant upon the procuring of sea-water, from the coast or from the ocean, I had long ago felt to be a great difficulty in the way of a general adoption of the Marine Aquarium. Even in London it is an awkward and precarious matter; how much more in inland towns and country places, where it must always prove not only a hindrance, but to the many an insuperable objection. The thought had occurred to me, that, as the constituents of sea-water are known, it might be practicable to manufacture it; since all that seemed necessary was to bring together the salts in proper proportion, and add pure water till the solution was of the proper specific gravity. . . .

" I took Schweitzer's analysis; but, as I found that there was some slight difference between his and Laurent's, I concluded that a very minute

accuracy was not indispensable. Schweitzer gives
the following analysis of 1000 grains of sea-water
taken off Brighton :—

Water	964·744
Chloride of sodium · .	27·059
Chloride of magnesium .	3·666
Chloride of potassium . .	0·765
Bromide of magnesium .	0·029
Sulphate of magnesia . .	2·295
Sulphate of lime	1·407
Carbonate of lime . . .	0·033
	999·998

"The bromide of magnesium and the carbonate
of lime I thought I might neglect, from the minute-
ness of their quantities; as also because the former
was not found at all by M. Laurent in the water
of the Mediterranean; and the latter might be
found in sufficient abundance in the fragments of
shell, coral, and calcareous algæ, thrown in to
make the bottom of the Aquarium. The sulphate
of lime (plaster of Paris) also I ventured to elimi-
nate, on account of its extreme insolubility, and
because M. Laurent finds it in excessively minute
quantity. The component salts were then reduced
to four, which I used in the following quantities :—

Common table salt	3½ ounces	}	Avoird.
Epsom salts	¼ „		
Chloride of magnesium . .	200 grains	}	Troy.
Chloride of potassium . .	40 „		

To these salts, thrown into a jar, a little less than
four quarts of water (New River) were added, so
that the solution was of that density that a specific
gravity bubble 1026 would just sink in it.
" The cost of the substances was—sulph. mag.
1d.; chloride mag. 3d.; chlor. pot. 1¼d.; salt, nil;

—total, $5\frac{1}{4}d$. per gallon. Of course if a larger quantity were made the cost of the materials would be diminished, so that we may set down 5*d*. per gallon as the maximum cost of sea-water thus made.* The trouble is nothing, and no professional skill is requisite.

"My manufacture was made on the 21st of April, 1854. The following day I poured off about half of the quantity made (filtering it through a sponge in a glass funnel) into a confectioner's show-glass. I put in a bottom of small shore-pebbles, well washed in fresh water, and one or two fragments of stone with fronds of green sea-weed (*Ulva latissima*) growing thereon. I would not at once venture upon the admission of animals, as I wished the water to be first somewhat impregnated with the scattered spores of the *Ulva;* and I thought that if any subtle elements were thrown off from growing vegetables, the water should have the advantage of it, before the entrance of animal life. This, too, is the order of nature; plants first, then animals.

"A coating of the green spores was soon deposited on the sides of the glass, and bubbles of oxygen were copiously thrown off every day under the excitement of the sun's light. After a week, therefore, I ventured to put in animals as follows:—

2	*Actinia mesembryanthemum.*	*Coryne ramosa.*
7	*Serpula triquetra.*	*Crisia eburnea.*
3	*Balanus balanoides.*	———*aculeata.*
2	*Sabella*——*?*	*Cellepora pumicosa.*
2	*Sabellaria (alveolata ?)*	*Cellularia ciliata.*
2	*Spio vulgaris.*	*Bowerbankia imbricata.*
1	*Cynthia (quadrangularis ?)*	*Pedicellina Belgica.*

* This was considerably over-rated: the cost is probably about $3\frac{1}{2}d$. per gallon.

" These throve and flourished from day to day, manifesting the highest health and vigour; the plants (including one or two Red Weeds that were introduced with the animals) looked well, and the water continued brilliantly crystalline. Within the succeeding month, specimens of *Actinia mesembryanthemum*, *A. anguicoma* and *A. clavata*, a *Trochus umbilicatus*, and a *Littorina littorea*, were at different times added.

" Six weeks have now elapsed since the introduction of the animals. I have just carefully searched over the jar, as well as I could do it without disturbing the contents. I find every one of the species and specimens mentioned above, all in high health; with the exception of some of the Polyzoa, viz. *Crisia aculeata*, *Cellepora pumicosa*, *Cellularia ciliata* and *Pedicellina Belgica*. These I cannot find, and I therefore conclude that they have died out; though if I choose to disturb the stones and weeds, I might possibly detect them. These trifling defalcations do in no wise interfere with the conclusion, that the experiment of manufacturing sea-water for the Aquarium has been perfectly successful.

" P. H. Gosse.

" 58, Huntingdon Street, Barnsbury Park,
     June 9, 1854."

The small quantity of water thus experimented upon remains to this time (April, 1856), having supported animal and vegetable life ever since without interruption, a period of two years. It is as transparent as the day it was put in, rivalling the water of the clearest rock-pool, from which it can in no respect be distinguished, either in its sensible qualities, or in its fitness for plants and animals. Since that time I have made other and larger quantities, with the same success; so that I can confidently recommend the formula for general adoption. The salts are sold in packets, with all needful directions, by Mr. Bolton, a chymist in Holborn.

It is interesting to find that the more subtile constituents of sea-water, as *Lime, Iron, Silica,* and *Iodine,* which I neglect in my formula, are gradually communicated to the artificial composition by use. Dr. George Wilson, of Edinburgh, who has analyzed portions of each of my preparations, after several months' use, finds traces therein of all these substances.

It is scarcely necessary to add, that, if you can conveniently procure water from the sea, you should do so by preference; " *si non, his utere mecum.*"

THE STOCK.

As I shall presently give some instructions concerning the modes of collecting both plants and animals, a few preliminary observations are all that will be needful here.

PLANTS.—What are the most suitable plants for an Aquarium? Not the Oar-weeds or Tangles (*Laminaria*); for though young specimens have an attractive appearance, they will not live long in captivity; they presently begin to decay, and slough off in slimy membranous shreds, filthy to look at, and hurtful to the living creatures. The *Fuci* live pretty well, but their sliminess and ugliness are fatal to their pretensions. From the

Red and the Green orders we must make our selection.

Of the former these will be found good. *Rhytiphlæa pinastroides*, the *Polysiphoniæ*, *Corallina officinalis*, *Delesseria alata*, *Chondrus crispus*, *Phyllophora rubens* (this, especially when dredged from deep water, is one of the very best), the *Griffithsiæ*, and some of the *Callithamnia*.

Of the Green weeds *Codium tomentosum* does pretty well, and affords food for some Mollusca that will eat nothing else. The *Cladophoræ* are good; *Bryopsis plumosa*, a most elegant little plant, flourishes in confinement; but the *Enteromorphæ* and *Ulvæ* are probably the best of all sea-weeds for our purpose, and the most easily procured on every shore.

The pieces of rock to which the plants are attached should be as clean as possible. All adhering sponges, in particular, should be carefully scraped off, unless they are wanted for immediate examination; as they are almost sure to die, and corrupt the soil and water with sulphuretted hydrogen, a most nauseous and noxious gas, which turns everything black with which it comes into contact.

ANIMALS.—Of the animals which thrive best in an Aquarium (speaking, of course, only from my own limited experience and observation) the following may be mentioned :—

FISHES. — The smaller Sticklebacks; young specimens of the Grey Mullet, which have lived for more than three years in the Zoological Society's Aquarium; the Blennies and Gobies; the Spotted Gunnel; the smaller Wrasses; the Rocklings; the Flounder; the Dab; the Eels.

MOLLUSCA.—The Sea-hare; the Periwinkle; the

commoner Tops; the Purple; the Murex; the Chitons; the Bullas; the Scallops; the Mussel; the Modioles; the Anomia; the Oyster; and some of the sand-burrowing bivalves, as *Venus, Mactra, Pullastra,* &c. *Gastrochæna* and *Saxicava,* burrowers in stone, may be readily kept, and are very interesting, especially the former, which I have had in confinement for many months, in more than a single instance, and still possess.

CIRRIPEDES.—The Acorn-barnacles (*Balanus* and *Chthamalus*), and the interesting little *Pyrgoma,* which is invariably found cemented to the plates of our larger Madrepore.

CRUSTACEA.—The Strawberry Crab, the Swimming Crabs, the Shore Crab, the *Ebalia,* the Masked Crab, the Soldier Crabs, the Broad-clawed Crab, the Shrimps, the true Prawns, the *Athanas ;* many of the *Entomostraca.*

ANNELIDES.—The Gold-comb; the Sabellas; the Serpulas; the Sea Leech; the Long Worm; the Terebellas.

ZOOPHYTES.—Many species of Sea Anemone, (except the Thick-horn, *Bunodes crassicornis,* which is precarious); both species of Madrepore.

The following are interesting, and may be preserved for a considerable time, but are rather more uncertain. Among FISHES, the Sea-scorpion (*Cottus*); the 15-spined Stickleback; the Butterfly Blenny; the Suckers; the Pipe-fishes. Among MOLLUSCA, all the Nudibranch, and most of the Tectibranch species; the *Naticæ,* the Cowry, the Whelk; the little *Rissoæ ;* the *Phasianella ;* the Cup-and-Saucer (*Calyptræa*); the lovely little *Acmea ;* many Bivalves; the *Cynthiæ,* and *Ascidiæ.* Among CRUSTACEA, the *Pisæ ;* the *Portuni ;* small specimens of the Common Crab and Lobster; the *Hippolytes; Pandalus; Gammarus; Idotea.* Among ANNELIDES, the Sea Mouse; the Nereides; and the *Planariæ.* Of ECHINODERMS, the *Cribella, Palmipes, Asterina, Asterias, Echinus* and *Cucumaria.*

PROCURING SPECIMENS.—By far the most interesting mode of acquiring your stock, is the collection of them by your own personal research. But as this is not in every case practicable, we must have recourse to the labours of others. In London, Mr. W. A. Lloyd, 19 and 20, Portland Road, New Road, is a "Dealer in Living Marine Animals, Sea-weeds, Artificial Sea-water, and Marine and Fresh-water Aquaria:" he will undertake the whole labour of supplying and stocking these interesting repositories of ocean life.*

TRANSMISSION OF SPECIMENS.—Both plants and animals should be forwarded to their destination as soon after they are collected as possible; but, if they are detained, they may be kept in pans of sea-water, exposed to the light. The vessels, however, must be protected from heavy rains, as the admixture of a large quantity of fresh water would be fatal to both plants and animals. Should much rain have fallen on a vessel containing specimens, it should be carefully tilted, so as to allow the fresh water, which, from its less specific gravity, will be lying on the surface, to run off without mingling with the other. If this be well done,

---

* Mr. Lloyd is *constantly* supplied with marine animals from the Kent, Dorset, South Devon, North Devon, and Welsh coasts, and *occasionally* from Cumberland, and the Channel Islands; so that his stock in London possesses a variety not to be found in any single locality on our shores.

most of the collection, at least that portion of it
which was nearest the bottom, may be preserved.

Living sea-weeds may be transmitted to long
distances without water. I used to employ a tin
box, enclosed by a basket. At the bottom I placed
a layer of refuse weed, the common *Fucus serratus*,
freshly gathered, and quite wet. On this bed I
laid the growing specimens (arranging the pieces
of rock so as not to shake about and injure the
plants) until the box was nearly full; over all,
refuse weed was again laid, filling up all hollows,
and so pressing the whole when the box was shut,
as to prevent any motion of the stones. The specimens arrived in the best condition, even the delicate *Delesseriæ* being uninjured.

Many animals may be forwarded in the same
way. The Mollusca, many of the Echinodermata,
several of the Crustacea, and all the Actiniæ are
transmitted with more ease and less danger thus
than in water. A handful of loose weed, wet with
sea-water, to keep a moist atmosphere around them,
may be thrown into a canister or jar, and the
animals placed in among it. The vessel should
not be *filled*, nor should any pressure be allowed
on the animals; the weed too, though fresh, must
be plucked, as pieces of rock would be injurious to
the more tender animals.

Fishes, however, many Crustacea, most of the
Annelida, all Medusæ, and the more delicate Zoophytes, require to be sent in sea-water. I sometimes use wide-mouthed jars of stone-ware, with
water-tight screwed tops, several of which may be
packed in a hamper; at other times a large 12
gallon zinc pail, protected by a wicker case, with
a screw lid, of which the central part is perforated

with minute holes; at others, four small zinc cans, of square form, with perforate tops, fitted into an open box, like case-bottles in a wine-hamper. All of these modes answer well; I know not to which I should give the preference; except that for Fishes the large pail is decidedly the best. If heavy stones or oyster-shells, very rich in Zoophytes and Annelides, be required, a common cabbage-net may be suspended from the lid of the pail in mid-water; the stones or shells, being put into this net, will be kept from injuring themselves or their neighbours by banging about upon the bottom.

The more brief the period during which the specimens are *in transitu* the better. Hence they should be always forwarded *per mail train*, and either be received at the terminus by the owner, or else be directed—"To be forwarded immediately by special messenger." The additional expense of this precaution is very small, and it may preserve half the collection from death through long confinement.

The packages should be opened immediately on arrival; several bowls, pans, &c. should be ready, each half-filled with sea-water. The water in the vessels just received should be carefully dipped or poured off, and the specimens placed one by one in the bowls. Thus you will not only see which are alive and healthy, and which are sickly or dead; but the weeds, shells, &c. will be rinsed from the sediment, which has been abraded during the rattling of the specimens in travelling. The specimens can afterwards be deposited in the Aquarium, their permanent home.

Should any of the more delicate animals appear much exhausted, they may often be restored by a prompt aeration of the water around them. This is most readily effected by means of the Syringe, as I shall presently describe.

If you can so arrange matters, it will be a useful caution to allow your plants exclusive possession of the Tank for a week or two; not putting any animals in, until you see bubbles begin to form all over the sides, bottom, and rock-work, when the sun-light shines on them. This appearance will indicate a growth of incipient vegetation, which will greatly lessen the chance of death when the animals are introduced.

Finally, be moderate in your desire of dominion. Do not overcrowd your Tank. It is far better to have it but half occupied at first, and to add to its population from time to time, than, by a too eager desire to see it filled, make it a Black Hole of Calcutta, and mourn over a host of corpses, the wreck, perhaps, of a single night. Half-a-dozen animals, averaging the bulk of a periwinkle, or a moderate-sized Sea-anemone, to every gallon of water, are quite enough to begin with.

GENERAL DIRECTIONS.

THE Aquarium is then established. The water, which at first is somewhat turbid, becomes in the course of a day or two clear and crystalline; the plants expand their feathery tufts in beauty, and the animals begin to take possession of their holes and corners, and to find themselves at home. But you must lay your account with the loss of some specimens; some will certainly die in the course of the first twenty-four hours, others in the first week. But those which survive the first ten days may be considered as pretty well established.

It is during this period that the grand trial of the experiment usually occurs. There is generally a large amount of animal matter attached to the sea-weeds, shells, and stones, which are received from the sea, such as minute Annelida, Mollusca, and Zoophytes : very many of these creatures are already dead, or die immediately; but being too minute to be detected and removed in detail, they decay, and presently contaminate the water. The first symptom of this is a slight dimming of the crystal translucency, which if unchecked soon increases to a milky whiteness, accompanied by a fetid odour, and terminates in the death of the whole animal collection.

PURIFICATION.—As soon as this begins to be perceived, the whole water should be drawn off by means of a siphon, without disturbing the sediment, into pans, into which, for the present, the plants and animals may be put. The Tank should be wiped out and rinsed, and then the water should

be filtered back into it. This is a very simple
process: a funnel (if of glass, earthenware, or
gutta-percha, the better) is placed over the tank,
with a bit of sponge pushed lightly into the top of
the tube, so as to allow the water to run through
in a narrow, thread-like stream. Replace the
plants and animals, reserving those pieces of rock,
or those shells, that look suspicious, which may be
kept in a bowl of water by themselves for a few
days, till their state appears more fully.

This process of bringing every drop of the water
into contact with the atmosphere, is an effectual
remedy for destroying the tendency to putrefaction ; as the animal fluids and solids held in suspension enter into combination with the oxygen of
the air, and form the pure innocuous gas called
ozone. The result will be that the milkiness will
rapidly disappear ; the water will assume a transparent clearness, which will in all probability be
permanent ; the plants will thrive, and the animals will be lively. This result will be rendered
still more secure by filtering the water through
pounded charcoal, and by allowing some pieces of
the same substance to float in the Tank.

Should it happen that, from oversight, or
ignorance, or any other cause, putrefaction has
thoroughly set in, still you need not lose the water.
Take out any animals that are yet alive, as well as
the weeds, and then put the water in *an open vessel*,
in some shed or out-house, where the fetor will be
of no consequence. Leave it there for the putre-
factive process to run its course, which it will do in
two or three weeks if the weather be warm. The
water will then gradually resume its original clearness and purity ; the weeds may now be replaced

in it; it should be well agitated frequently, and soon you may put in, cautiously at first, your animals, and every thing will be right again.

OCCASIONAL DEATH.—It will still be needful to exercise a watchful supervision of the collection. It must be remembered that both the animals and plants are not in their natural circumstances, and that a certain amount of violence is done to their habits. Death, which spares them not at the bottom of the sea, will visit them in the Aquarium; and hence the vessel should be occasionally looked over, *searched*, as it were, to see if there be any of the specimens dead. If the plants show an orange hue in patches, they must be taken up, and the diseased parts cut clean away. Dead animals must be at once removed, or contamination will soon result. The eye will soon recognise the individuals, and will miss the familiar forms; but you must not too hastily conclude that an animal, which you have been accustomed to see playing about, is dead, because you have not observed it for some days, and cannot find it. Probably it has secreted itself in some corner or crevice, whence it will emerge in a day or two. Still such a circumstance should excite your vigilance.

INSTRUMENTS.—For removing dead specimens or the like, a pewter spoon bent up to a right angle, with the shaft tied to a slender stick, is very useful. You can, if you please, make a more elegant affair of it. Two or three simple sticks or rods, some of them widened, spade-like, at the end, are also useful for pushing the specimens to any required point. And one or two small nets made by stretching a bit of lace or muslin over a ring of wire, fastened to a rod, will serve to catch and lift out such animals as you wish to transfer, for examination, or any other purpose, to another vessel. As a general rule, however, they should be disturbed as little as possible, and never handled.

ARTIFICIAL AERATION.—Although living and healthy plants will educe and throw off, under the influence of light, oxygen, in sufficient quantity to maintain in health a given number of animals, yet the artificial admixture of atmospheric air with the water may be employed as a valuable auxiliary. I have used it with marked benefit; often having revived animals thereby, which, from the exhaustion of the water, were apparently in a dying state. Its utility as a means of maintaining the purity of the water is still more obvious; since it is by the frequent and successive presentation of the particles of water to the air, that the animal excretions which they hold in suspension, become chemically changed, and deprived of their putrescent qualities. This is what takes place in nature. By the perpetual dashing of the waves against the shore, and especially against the ragged rocks, an immense quantity of air becomes entangled, in the form of minute bubbles, which by the various currents are diffused through the sea, and even carried to considerable depths, before they rise to the surface and become dissipated. Thus the violent agitation of the sea is a powerful agent in its purification.

One of the simplest modes by which this object can be effected, is the drip-glass. I have been accustomed to suspend over the Aquarium, a perforated bell-glass (I think it is called a bee-glass) of suitable size, into the orifice of which a bit of sponge may be pushed, or a cork drilled with small holes. The cord which suspends the drip-glass passes over a pulley at the top of the window, so as to be raised or lowered at pleasure. Every morning sufficient water from the Tank is drawn or dipped off, to fill the drip-glass, which is then hoisted to its full height. The contents run out in slender steams, or in a rapid succession of drops, which, passing through some four or five feet of air, before they reach the Tank, become effectually purified.

A convenient mode of aeration is that effected by strongly syringing the water. The instrument should be at least $1\frac{1}{2}$ inch in diameter, and should be raised above the surface at every down-stroke. After a few moments' work, the whole Tank will be quite white with minute bubbles of air, resembling the sea when the waves dash and boil among the rocks.

The same purpose may be more efficiently accomplished at a slight expense, in a manner which would greatly augment the elegance of the Aquarium. In the engraving placed at the commencement of this treatise, I have represented a Fountain-Aquarium,—a form of the invention particularly suitable for a conservatory or hall. It needs but a vessel fixed, as a reservoir, at some distance above the level of the Tank, in a higher story for example, whence a supplying tube may descend, and passing beneath the floor, ascend through the foot of the vase, to the surface of the water. All the visible portion may be easily concealed among the rock-work; while from the extremity a jet would play, proportioned in force to the weight of the supplying column, or, in other words, to the height

of the reservoir above the surface.   It would be
needful to make the apparatus of some incorrodible
material;—gutta percha, for instance, for the tube,
with a nozzle of glass;—as metals would be acted
on by the sea-water, and form noxious oxides. The
water might either be carried up to the reservoir, or
pumped up by an obvious extension of the apparatus.

Such a modification would doubtless be as
efficient as it would be elegant.   The constant, or
at least frequent, dissemination of the water through
the air would keep the whole volume in agreeable
coolness, as well as maintain its sparkling clearness and purity.

In a well-regulated Tank, however, none of
these modes are necessary.   My oldest reservoir,
which has been in constant occupation for more
than two years, never has any artificial aeration,
except an occasional syringing, and that is often
intermitted for months together.   The surface is
now and then agitated with a stick, and broken
by the addition of fresh water to supply the loss
by evaporation, and this is all the external aid it
receives.   *Yet the water generally maintains the
most crystalline transparency and purity.*

EVAPORATION.—If the Tank remain habitually
uncovered, or protected only by a coverlid of muslin, daily evaporation will soon reduce the volume
of the water, and increase its specific gravity.   The
pure water alone rises in vapour, the various salts
held in solution remaining the same in quantity,
though the water should be reduced to half its
original bulk.   It is therefore needful that additions
of pure *fresh* water (not *sea*-water) be made from
time to time, to replace the loss by evaporation.
Distilled water is of course the best, but, practically,

river-water will answer perfectly well. The time and quantity of these additions ought to be regulated by a hygrometer, the specific gravity of the sea-water being maintained at about 1027, which is the average density of the waters of the Atlantic. A tolerable approximation to accuracy, however, may be made, by marking on the vessel the surface-level at first, and always maintaining the same level. A glass cover greatly prevents loss from evaporation, as will be manifest by the condensed moisture on it, especially after a cold night.

CLEANSING THE SIDES.—The sea-water constantly holds in suspension millions of the spores (or seeds) of Algæ, ready to adhere and grow as soon as they find a resting-place, and these are particularly abundant in the warm season. Whether those of the green kinds, the *Chlorosperms*, such as the *Ulva, Enteromorpha*, and various kinds of *Confervæ*, be more plentiful than others, or whether they are more easily satisfied with a place congenial to their growth, I know not; but these grow most obviously, in the proportion of a thousand to one. Before we have kept our tank stocked a fortnight, its transparent sides begin to be sensibly dimmed, and a green scurf is seen covering them from the bottom to the water's surface, which constantly accumulates, soon concealing the contents of the vessel from distinct observation. On examining this substance with a lens, we find it composed of myriads of tiny plants, mostly consisting of a single row of cells of a light green hue, forming minute threads which increase in length at the extremity; others display small irregularly puckered leaves of deeper green, which develop themselves into *Ulvæ*, or *Enteromorphæ*.

If we design the Aquarium to be of any service to us in the observation of its contents, this growth must be got rid of, or we might as well have a vessel with opaque sides. Here then we bring in the aid of the Periwinkle, which may be bought alive of any London fishmonger, half-a-pint for a penny. Exclusively a vegetable-eater, he delights in the green sea-weed, and nothing can be more congenial to his palate than these tender succulent growths. The little Yellow Winkle, so abundant on weedy rocks, possesses a similar appetite; but he is less suitable for the service required, inasmuch as his constitution appears unable to bear constant submersion; his habit is to live a good deal exposed to the air, and even to the hot sun, and this seems essential to his health. I have found that if this little species be collected, pretty as the individuals are, they crawl around the sides for a day or two, as if seeking a more genial dwelling, and then one by one fall to the bottom and die. There is, however, another genus of Univalve Mollusca which may be made equally available with the Periwinkle, if indeed it be not superior for the purpose. I allude to those evenly conical shells, which belong to the genus *Trochus*, sometimes called from their form, Tops. Two species, *T. cinerarius* and *T. umbilicatus*, are scarcely less abundant on our weedy shores than the Periwinkles; the former of a dull purplish-grey, marked with close-set zigzag lines; the latter rather flatter, usually worn at the summit, of a dull olive or green, with narrow reddish bands radiating from the centre. Both are pearly in the interior, but the latter species is brilliantly irridescent.

These Tops and the common Periwinkle are very

useful inhabitants of a marine tank; they make themselves at home, and feed readily. It is interesting to watch the business-like way in which they proceed. At very regular intervals, the proboscis, a tube with thick fleshy walls, is rapidly turned inside out to a certain extent, until a surface is brought into contact with the glass, having a silky lustre; this is the tongue; it is moved with a short sweep, and then the tubular proboscis infolds its walls again, the tongue disappearing, and every filament of conferva being carried up into the interior from the little area which had been swept. The next instant, the foot meanwhile having made a small advance, the proboscis unfolds again, the tongue makes another sweep, and again the whole is withdrawn; and this proceeds with great regularity. I can compare the action to nothing so well as to the manner in which the tongue of an ox licks up the grass of the field, or to the action of a mower cutting down swathe after swathe as he marches along. The latter comparison is more striking for the marks of progress which each operator leaves behind him. Though the confervoid plants are swept off by the tongue of the Mollusk, it is not done so cleanly but that a mark is left where they grew; and from its peculiar form and structure, the tongue leaves a series of successive curves all along the course which the Mollusk has followed, very closely like those which mark the individual swathes cut by the mower in his course through the field.

The unsightly appearance thus left must be got rid of by mechanical means. A stick with a bit of rag tied around its end, or, what is better still, a brush made on purpose, like a nail-brush with a

very long handle proceeding from the side—a sort
of hearth-brush in miniature, fitted with very close
and stiff bristles—will rub off the greenness. It
may be used about once in a month, or oftener in
summer. On the stones of the bottom, the cement
and rock-work, and even on that side of the tank
which, being next to the window, is not used for
observation, I would recommend that the green
growth be not interfered with, but that the marine
plants be allowed to grow undisturbed. A crop of
self-sown weeds in the tank is far more valuable
than such as have been introduced on loose stones.
And even from very early age the green growth is
found to throw off a copious supply of oxygen
bubbles. Care, too, must be taken not to molest
or annoy the animals needlessly, as also to leave
undisturbed any masses of spawn that may have
been deposited on the glass.

TURBIDITY.—Occasionally the water in a tank,
which has hitherto been quite translucent, becomes
all on a sudden so turbid as completely to hide the
contents from view, except such as are close to the
glass. This turbidity may arise from either of two
causes. If it is of a grey or whitish hue, forming
clouds here and there, which disperse and form
again elsewhere, the microscope will show that it is
composed of an innumerable multitude of animalcules belonging to the Class INFUSORIA. Their
presence is not an evil, but rather a means whereby
an already existing evil may be remedied. Their
sudden increase to such an extent as to be thus
appreciable to the senses, is symptomatic of organic
matters in the tank in a state of decomposition. If
we allow a minute worm that has died to lie at the
bottom of the Tank, we shall see in a day or two,

if we watch it with a powerful lens, that it is encompassed by a little cloud of moving atoms, which are the animalcules in question, and which are busily engaged in devouring not only the solid parts, but also the juices and invisibly minute particles that float off; and thus in a very short time they effectually dispose of the offensive substance. So, in the case of their increase to the extent just supposed, of producing a general turbidity, they will, if left to themselves, soon clear away the decomposing matter if it be not too great, and then themselves gradually disappear, allowing the water to resume its original clearness. As soon, however, as we perceive such an appearance, we should carefully transfer the principal animals to another vessel, and search for the decomposing bodies, on the removal of which the water will presently be transparent and sweet as before.

But the opacity of the water may be dependent on a totally different cause. If it is of a green colour, rapidly deepening in intensity, it is vegetable in its origin, and arises from an infinite number of the spores (or seeds) of green *Algæ* dispersed through the fluid, and held in suspension there. Now, this appears to have no deleterious influence either on the plants or animals, which live and thrive as well as when the water is clear; but it is annoying because of its unsightliness, and because it effectually interferes with our observation of our cherished favourites. It is, too, a most inveterate evil; unlike the former, it is not self-curative, at least not certainly so, and it cannot be foreseen. I have had a large vessel that had been in full occupation for a year and a half,—during the whole of which time it had remained brilliantly colourless, — suddenly, without any imaginable cause, become green; and in the course of two days be so opaque that objects could not be discerned an inch from the sides.

The lens will not detect anything in the fluid in this case; it requires a very high power of the compound microscope to resolve the cause. With a magnifying power of 560 diameters, we see an immense number of oval atoms, apparently colourless (but doubtless, having a very slight tinge of green visible only in the aggregate), and not more than $\frac{1}{5000}$ of an inch in diameter. These I conceive to be the spores of some green *Oscillatoria*, or some kindred plant; for there is a tendency to the accumulation of the films of such plants in the vessels in which the phenomenon exists.

Sometimes this evil will continue unchanged for many months, and then clear away as suddenly as it came. At others, it will diminish and promise a return of transparency, then suddenly return, and set in as dark as before.

Mr. W. A. Lloyd has succeeded in overcoming this difficulty. By drawing off the green water, and putting it into a dark closet, he finds that in two or three weeks the turbidity quite disappears, the water resuming its pristine transparency. The explanation is doubtless as follows : light is necessary to the life of plants, or at least the green colouring principle in them cannot be developed without light; if, then, this be denied, the plants must wane and die. Now the opacity, as I have intimated, consists of the living germs of green plants ; and these on being deprived of light gradually die away ; after which the water is quite fit for use again. I have myself instituted experiments on the subject; and so far as I have proceeded, my results agree with those of Mr. Lloyd; except that I have found a seclusion of several months necessary. This was, however, in the winter.

Cloudiness may often be removed by the action of living animals. The Bivalve Mollusca collect the organic particles that float in the water, as the currents pass over their capacious gills, and either feed on them, or consolidate them into cylindrical rolls, which fall to the bottom, enveloped in mucus. Hence they are of great service in making turbid water limpid and bright. An oyster or two, according to the size of the vessel, will answer exceedingly well, as they are very hardy, and can be procured alive anywhere, and almost at all seasons.

FOOD.—I am continually asked, *how, with what,* and *how often* I feed my animals. My invariable reply is,—"Not at all." I do not find that they need any supply but what they procure for themselves. In a well-stocked and established Tank, the vegetable-feeders find a sufficient pabulum in the ever-growing weed; and all the carnivorous species are maintained in condition by the hosts of INFUSORIA and ENTOMOSTRACA that are always swarming. The lens shows these tiny creatures to be abundant in every collection of water that has been kept for a short time; and, as they breed very fast, their increase is sufficient to meet the demands of their superiors in organization. At least, I find that my well-filled Tanks need no other supply than this, and what the more predatory kinds occasionally obtain by the death of their fellows. For amusement, indeed, the Actiniæ, the Madre-pores, the Prawns, and the Crabs, may be fed;

and then the best diet is the lean of raw meat,
cut into minute fragments; but it should be very
sparingly done, and the rejected atoms carefully
removed, lest putrefaction set in and spoil the
whole.

INSTRUCTIONS FOR COLLECTING.

TIME.—What is commonly called *low-water*—
that is, the time when the ebbing-tide recedes to
the utmost point—is the period to be chosen for
shore-collecting, as comparatively few marine animals or plants habitually live in situations where
they are long exposed to the air and sun.   But the
lower the level, the more rich becomes the harvest;
and hence the time of spring-tide is the most productive, when the recess of the tide is the greatest.
Spring-tides occur twice every month, viz. about
the time of new and full moon; the very best tides
of all are those of the second day after the change
of the moon; but for two days before and two days
after that, they recede very far; so that we may
consider those weeks which commence two days
before the change of moon and end five days after
it, as good collecting periods, while the alternate
weeks are nearly useless.   The full-moon tides are
generally greater than the new-moon; and those
about the time of the equinoxes, or the spring-tides
of March and April, and of September and October,
are the best of the whole year.   Prevailing winds,
however, exercise some influence on the amount of
recess of the tide.

The time of lowest water on any particular day
can be readily ascertained from local tide-tables;

and the young collector, in choosing the locality of his operations, will pay attention to this point beforehand, that he may select a place where the time of low-water on the days of spring-tides is the most convenient for his occupation. For instance, the time of low-tide on the day of full-moon is about noon on our western shores, but about six o'clock on the Sussex coast.

IMPLEMENTS.—I use a wicker-basket with a flat bottom and straight sides, divided into compartments. In two of these fit wide-mouthed jars, such, for example, as are used for preserves: if made of glass they are the better, as admitting a more ready examination of their contents; but jars of white ware or stone-ware will do. The larger objects procured are put into these, but I commonly carry also a wide-mouthed phial, such as the chemists keep *quinine* in, fitted into a third compartment, to receive the minuter and more delicate things. Then there is a fourth division running the whole length of the basket, in which lie a hammer and chisel, and which may receive large shells, crabs, &c. that do not require constant immersion. A geologist's *hammer* with a cutting edge, as well as a striking face, is the most useful; and the *chisel* must not be such as carpenters use, but one made wholly of iron, tipped with steel, such as is used by smiths, and technically called a *cold chisel*.

Sometimes, especially if the shore we are about to search be strewn with large stones or boulders, it will be well to secure the attendance of a man with a crowbar, to turn over the stones, as on their under surface, and beneath their shadow, valuable specimens are often found. With the same instrument, inserted into the fissures, great pieces of loose slaty rock may be wrenched off, which are very productive.

COLLECTING SEA-WEEDS.—Thus armed, we sally forth, choosing for our explorations a spot where low dark ledges of shelving rock run out into the sea, full of clefts and fissures half concealed by Bladder-weed and Tangle; or where the solid rock shoots up in irregular angular masses, scooped and hollowed into numberless little pools and basins, with dark, slimy caverns here and there, and rifts of shingly sand between. An unpractised foot would find the walking precarious and dangerous, for the rocks are rough and sharp; and the dense matting of black Bladder-weed with which they are covered, conceals many abrupt and deep clefts beneath its slimy drapery. These fissures, however, are valuable to us. We lift up the hanging mass of olive-weed (*Fucus*) from the edge, and find the sides of the clefts often fringed with the most delicate and lovely forms of sea-weed; such, for example, as the winged Delesseria (*D. alata*) which grows in thin, much-cut leaves of the richest crimson hue, and the feathery Ptilota (*P. plumosa*) of a duller red. Beneath the shadow of the coarser weeds, as well as in open pools, delights also to grow the *Chondrus*, in the form of little leafy bushes, each leaf widening to a flattened tip. When viewed growing in its native element, this plant is particularly beautiful; for its numerous leaves grow with refulgent reflections of azure, resembling the colour of tempered steel. This weed when dried is useful for making jellies, and constitutes the Carrageen Moss of the shops.

We may observe among the sea-weeds many tufts of a small species, whose leaves are much and deeply cut, with the divisions rounded, and the general outline of the leaf pointed. Some specimens are of a dull purple, others of a rich yellow hue; and I refer to the species as an interesting example of the influence of light on the colour of marine plants. The yellow specimens are exposed to the sun's rays; the purple ones are such as have grown in deep shadow. The species is the *Laurencia pinnatifida* of botanists.

Turning from the hidden clefts, we explore the deep pools that lie between the ledges. High wading-boots are necessary for this purpose, as we have to work in the water. The great Oar-weeds and Tangles (*Laminaria*) are growing here, large olive sea-weeds that wave to and fro with the undulations of the sea; the former a long narrow puckered frond of brown colour; the latter, a broad smooth leathery expanse of deeper colour on a slender stalk, splitting with age into a number of lengthened fingers or ribbons, and hence called the Fingered Tangle (*Laminaria digitata*). Among these grow clusters of an elegantly frilled species, of delicate thin texture, and yellow-brown hue, bearing no slight resemblance to the tresses of some fair lady: this also is a *Laminaria*, but I am not quite sure whether it is the young state of the former species, or entitled to a name of its own. In the latter case, it is the *L. phyllitis* of botanists.

In these deep pools grow also many bunches of broad dark-red leaves, generally about as large as one's hand, smooth and glossy, of a dark crimson hue, but apt to run off into a pale greenish tint towards the tips; their edges have often little leaves growing on them. It is the Dulse or Dillis

(*Rhodymenia palmata*), which is eaten by the poor of our northern shores as a luxury.

This is a showy plant, very beautiful when its tufts of large deep-red fronds are seen in the sea, where the perpetual wash of the waves keeps their surface clean and glossy, but not very suitable for an Aquarium. Its leaves soon decay; spots of orange-colour begin speedily to appear, which increase fast, and, uniting into large patches, slough off in slimy shreds. The appearance of an orange-colour, on crimson or purple weeds, is always a sign of the death of that part, and is the infallible precursor of decay. As soon as it appears, or at least if it begins to increase, the specimen should be ejected without mercy; as the diffusion of the gases from decaying vegetable matter is speedily fatal to most animals.

In deep pools, and narrow clefts near the verge of lowest water, where the overshadowing rock excludes the sun's rays and imparts a genial obscurity, grow several of our most delicate and beautiful *Algæ*. Foremost among them is the Oak-leaved Delesseria (*D. sanguinea*), with tufts of crimson leaves, exquisitely thin, much puckered at the edge, and strongly nerved. The *Iridæa*, whose leaves are smooth and leathery, and of a dark brownish scarlet, is often the companion of the former. Here, too, we find the *Phyllophora*, another weed of brilliant red hue, with unnerved leaves much divided, giving origin to other leaves, and these again to others. It is usually much covered with the cells and shrubs of various species of *Polyzoa*, exquisitely beautiful objects for the microscope. The *Gelidium corneum* is another fine red weed, commonly of small size and slender, but prettily fringed with

processes all round the edges of the leaves. This
and the preceding are very hardy in confinement,
and form very suitable plants for an Aquarium.

When we can no longer work at so low a level,
we recede to the slopes of the ledges yet uncovered,
and find other species in the quiet sheltered pools.
A weed is found here, growing in dense mossy
patches on the perpendicular and overshadowed
edges of the rock, which, when examined, looks
like a multitude of tiny oval bladders of red-wine,
set end to end in chains. This pretty sea-weed is
called *Chylocladia articulata*.

Here also grows the stony Coralline, a plant
bearing some resemblance to that just named, in
the peculiar jointed form of its growth. Low-lying
pools are often incrusted with a coat of stony or
shelly substance of a dull purple hue, having an
appearance closely like that of some lichens; the
crust investing the surface of the rock, and adhering
firmly to it, in irregular patches, which continually
increase from the circumference, in concentric zones.
This is the young state of the *Corallina officinalis*,
which by and by shoots up into little bushes of
many jointed twigs, diverging on every hand, or
hanging in tufts over the edges of the rock-pools.
Young collectors are eager, I perceive, to seize such
specimens as are purely white; but this condition
is that of death; in life and health, the shoots are
of the same pale purple hue as the lichenous crust.
This plant in both states (for plant it undoubtedly
is, though principally composed of lime, and of
stone-like hardness) is suitable for a tank, as it
survives and flourishes long; and your pieces of
rock-work you may select from such places as are
covered with the purple crust.

The most valuable plant of all for our purpose is the Sea Lettuce (*Ulva latissima*). Every one is familiar with its broad leaves of the most brilliant green, as thin as silver-paper, all puckered and folded at the edge, and generally torn and fretted into holes. It is abundant in the hollows of the rocks between tide-marks, extending and thriving even almost to the level of high water, and bearing with impunity the burning rays of the summer's sun, provided it be actually covered with a stratum of water, even though this be quite tepid. It therefore is more tolerant than usual of the limited space and profuse light of an Aquarium, where it will grow prosperously for years, giving out abundantly its bubbles of oxygen gas all day long. It is readily found ; but owing to the excessive slenderness of its attachment to the rock, and its great fragility, it is not one of the easiest to be obtained in an available state. The grass-like *Enteromorphæ* have the same qualities and habits, but their length and narrowness make them less elegant. The *Cladophoræ*, however, are desirable ; they are plants of very simple structure, consisting of jointed threads, which grow in dense brushes or tufts of various tints of green. Some of them are very brilliant: the commonest kind is *C. rupestris*, which is of a dark bluish-green ; it is abundant in most localities.

These are a few of the sorts of sea-plants which are met with in the situations I have described. In order to transfer them to an Aquarium, a portion of the rock on which they are growing must be removed. These plants have no proper roots, and therefore cannot be dug up and replanted like an orchis or a violet, but adhere by a minute disk to the surface of the rock, and, if forcibly detached, die. I therefore bring the hammer and chisel into requisition, and split off a considerable fragment of the solid stone, which then, with the plant adhering to it, is placed in the Aquarium. This is often a difficult, always a delicate operation; the rock is frequently so hard as to resist the action of the chisel, or breaks at the wrong place; sometimes, on the other hand, it is so soft and friable as to crumble away under the implement, leaving only the isolated plant deprived of its attachment; and sometimes at the first blow, the sea-weed flies off with the vibration of the shock. Often we have to work under water, where the force of the blows is weakened and almost rendered powerless by the density of the medium, and where it is next to impossible to see with sufficient clearness to direct the assault.

As the plants are detached, they are placed one by one in security. The finer and more delicate ones, as the *Delesseria* for instance, are immediately dropped into a jar of water; for only a few minutes' exposure of their lovely crimson fronds to the air, would turn them to that dull orange colour already mentioned as the sign of incipient decay. The hardier sorts are laid in the basket,—a layer of damp refuse-weed being first put in to receive them, —and covered lightly with damp weed. The degree of moisture thus secured is sufficient to preserve many species from injury for hours. Thus they are brought home.

COLLECTING ANIMALS.—I have been speaking of the haunts of the living *Algæ*, and of the manner of procuring them; because in sequence of idea these come first into consideration. But in point of fact, the search for animals goes on simultaneously with the process just described; the same haunts which are affected by the marine plants conceal various animals; and it is one of the great charms of collecting, that you never know what you may obtain at any moment. The expectation is always kept on the stretch : something new, or at least unthought of, frequently strikes the eye, and keeps the attention on the *qui vive.*

Close examination of the fissures of the pools, of the rough and corroded stones that have been fished up, and even of the sea-plants themselves, reveals many curious creatures of various kinds and forms, each of which, when found, is seized and consigned to one or other of the jars. The plants often bear the more delicate Zoophytes, as *Coryne, Sertularia, Campanularia,* &c. growing parasitically upon them; and some interesting Sponges, as *Grantia compressa* and *G. ciliata.* But more generally the Sponges are found incrusting the surface of the rocks in the darkest places, especially on the sides of caverns, intermixed with many species of the Polyzoa.

The Sea-anemones (*Actinia,* &c.) adhere to rocks; the common Smooth species (*A. mesembryanthemum*) often high up, exposed to the air; but the rarer kinds generally in the sheltered crannies and basins, in gravelly fissures, or on the under surface of stones. They must be carefully dislodged by inserting the finger-nail beneath the base, and gradually shoving them off; but those sorts that live in holes must be chiselled out.

Many of the Star-fishes, Sea-urchins, and Sea-cucumbers are to be procured by turning over loose stones at the lowest tide-level; various species of

Annelides haunt the same places, and some of these are of surpassing beauty. Many curious kinds of Crabs and other Crustacea, too, and in spring the elegant Nudibranch Mollusca, reward the labour of stone-turning.

The Univalve Mollusca crawl freely over the surface of the rocks, or roam amid the umbrageous foliage of the weeds that fringe the clear pools; whither also many of the lithe and slender Worms, and the swimming Crustacea, as the Prawns and their allies, resort. Some of the Bivalve Mollusca burrow into the solid rock itself, and the Acorn Barnacles are seated by thousands on its surface. Most of the burrowing Bivalves, however, are to be dug out of the sand or mud of the flat shore, and many interesting animals are found on the beds of sea-grass (*Zostera*) that grows on such a coast. These are collected by means of a keer-drag, a form of net which the reader may find described, with the mode of using it, in my "Aquarium," p. 56.

DREDGING.—To the same work I must refer for a description of dredging and its prolific results, whereby the bottom of the deep sea is scraped and the varied contents brought to light. Multitudes of animals of the highest interest are procured by dredging, that the shore-collector would never find; and yet shore-collecting must always be the main resource, at least of the majority.

TOWING.—One more means of obtaining animals remains to be mentioned,—the surface-net. This may be made of stout muslin, in the form of a bag, two feet deep, sewed on a thick brass ring a foot in diameter, which is screwed at the end of a staff six feet long. The staff should be of tough wood, such as hickory or lance-wood. The net is held at the surface of the sea, the collector sitting in a boat rowed gently along. The afternoon and evening of a calm sunny day is most productive, especially in the latter part of summer or autumn, when the lovely *Medusæ*, the little *Beroe*, and many forms of freely swimming ANNELIDA and CRUS-TACEA occur in abundance. At frequent intervals, the bag of the net must be reversed and plunged into a glass jar of clear water, when the captives will float off into the vessel.

For many details of this and other modes of collecting, and of the history of the curious creatures obtained thereby, I beg once more to refer to my "Devonshire Coast," "Aquarium," and "Tenby," which are expressly devoted to these subjects.

THE END.

MARINE
NATURAL HISTORY CLASS.

In the summer of 1855, I met at Ilfracombe, on the coast of North Devon, a small party of ladies and gentlemen, who formed themselves into a Class for the study of Marine Natural History. There was much to be done in the way of collecting, much to be learned in the way of study. Not a few species of interest, and some rarities, fell under our notice, scattered as we were over the rocks, and peeping into the pools, almost every day for a month. Then the prizes were to be brought home, and kept in little Aquariums for the study of their habits, their beauties to be investigated by the pocket-lens, and the minuter kinds to be examined under the microscope. An hour or two was spent on the shore every day on which the tide and the weather were suitable; and, when otherwise, the occupation was varied by an indoor's lesson, on identifying and comparing the characters of the animals obtained, the specimens themselves affording illustrations. Thus the two great desiderata of young naturalists were attained simultaneously; they learned at the same time how to collect, and how to determine the names and the zoological relations of the specimens when found.

A little also was effected in the way of dredging the sea-bottom, and in surface-fishing for Medusæ, &c.; but our chief attention was directed to shore-collecting. Altogether, the experiment was found so agreeable, that I propose to repeat it by forming a similar party every year, if spared, at some suitable part of the coast.

Such ladies or gentlemen as may wish to join the Class should give in their names to me, early in the summer; and any preliminary inquiries about plans, terms, &c. shall meet the requisite attention.

P. H. GOSSE.

58, Huntingdon Street,
Islington London.

PROPOSED WORK

ON THE

BRITISH SEA-ANEMONES.

Mr. Gosse has for some years been collecting materials for a complete History of our native Sea-Anemones, with illustrations of every species, drawn and coloured by himself from living specimens.

In order to further this project, he respectfully invites the co-operation of his kind scientific friends at various parts of the British and Irish Coasts, who may materially assist him by transmitting to him (*free of expense*) specimens of all species that are not common everywhere.

An Anemone of medium size may be safely sent *by post*, in a small tin-canister, *without water*, but with a small tuft of damp sea-weed to maintain a moist atmosphere around the animal. A piece of paper should be *pasted* round the canister, to secure it, and also to receive the address; and the whole would probably come within the weight covered by a twopenny or fourpenny stamp.

58, Huntingdon Street, Islington.

WORKS BY PHILIP HENRY GOSSE, F.R.S.

THE AQUARIUM;

AN UNVEILING OF

THE WONDERS OF THE DEEP SEA.

Post 8vo. with coloured and uncoloured Illustrations, 17s.

"Those who have had the gratification of spirit-companionship with Mr. Gosse in his former rambles, will rejoice to find themselves again by his side on the shores of Dorset. He has the art of throwing the 'purple light' of life over the marble form of science; and while satisfying the learned by illustrations and confirmations of what they knew before, he delights the seekers of knowledge, and even of amusement, by leading them into profitable and pleasant paths 'which they have not known' . . . . . . The volume ought to be upon the table of every intelligent sea-side visitor. It would be injustice to close these remarks without paying a tribute to the singular beauty, both of design and execution, of the plates which accompany the work."—*Globe*, June 22, 1854.

"To all who have looked with interest upon the collection of marine aquatic animals in the Zoological Gardens, and observed with attention their wondrous development of form and function, this book, by an eminent lover of Nature's marvels, will be a delightful and welcome companion. Mr. Gosse has himself dived into the bejewelled palaces which old Neptune has so long kept reluctantly under lock and key, and we find their treasures set before us with a freshness and fidelity which afford welcome and instructive lessons to naturalists of all ages. . . . It is a charming little volume, and an admirable pocket companion for visitors to the sea-side."—*Literary Gazette*, July 15, 1854.

"The beautiful little work now before us. . . . . Every page of this fascinating work is quotable. . . . A fitting ornament for the drawing-room table."—*Chambers's Journal*, Aug. 1854.

LONDON : JOHN VAN VOORST, PATERNOSTER ROW.

WORKS BY PHILIP HENRY GOSSE, F.R.S.

A MANUAL OF MARINE ZOOLOGY

FOR

THE BRITISH ISLES.

Two Vols. Foolscap 8vo. with nearly 700 Engravings, 15s.

THIS Work gives in plain English terms the characters by which to determine the Class, Order, Family, and Genus of *every animal* known to inhabit the British Seas. Every Genus is illustrated by a figure, drawn by the Author, principally from nature, and is accompanied by a list of the recognised Species.

A need long felt is supplied by this book, which, it is hoped, will be found a valuable *vade mecum*, if not indispensable, to every visitor to the sea-side, who desires acquaintance with its living treasures.

Every Class is introduced by a *résumé* of the most interesting points of its Natural History, with notes of the localities frequented by the Species, and directions for identifying them.

PART I.

I. SPONGES.	V. STARFISHES.	IX. CRUSTACEA.
II. FORAMINIFERA.	VI. TURBELLARIA.	X. CIRRIPEDIA.
III. ZOOPHYTES.	VII. ANNELIDA.	XI. MITES.
IV. MEDUSÆ.	VIII. ROTIFERA.	XII. INSECTS.

PART II.

XIII. POLYZOA.	XVI. BRACHIOPODA.	XIX. CEPHALOPODA.
XIV. TUNICATA.	XVII. PTEROPODA.	XX. FISHES.
XV. CONCHIFERA.	XVIII. GASTROPODA.	XXI. MAMMALIA.

LONDON: JOHN VAN VOORST, PATERNOSTER ROW.

WORKS BY PHILIP HENRY GOSSE, F.R.S.

Just Published,

TENBY:

A SEA-SIDE HOLIDAY.

With 24 Plates, coloured, post 8vo. 21s.

" Here we have another issue of the fertile pen of Mr. Gosse, and another of his delightful sea-side books. It is fully worthy of its predecessors in pleasant gossip, in interesting information, in important scientific novelty, and in variety and beauty of illustration."—*Athenæum*, May 31, 1856.

" It is the history of a month spent by a man of research, in the pursuit of a favourite study, under favourable circumstances; and is full of original investigations, successful observations, and pleasing descriptions of the impressions produced by novel objects upon an unaffected and healthy mind. It is a book we cannot read without regretting, as we pass from page to page with increasing interest, that we were not his companions. . . . . . . No intelligent reader can rise from the perusal of 'Tenby' without gaining much knowledge from a delightful book."—*Eclectic Review*, June, 1856.

" Mr. Gosse tells us how he got to Tenby, talks of the places there,—the caverns, Monkstone, North Cove, Hean Castle, Hoyle's Mouth, Tenby Head, and other places to be visited; shows where the marine animals, his favourites, most abound; teaches how to get at them, when to catch them in a visible condition, how to keep them, how to study them, and what their points of interest are. Of such matters is the book made up, and to us it seems to be perfect in its way."—*Gardener's Chronicle*, May 17, 1856.

" The natural history is admirable, the descriptions picturesque and vivid in a very uncommon degree, and the illustrations excellent. Mr. Gosse has, in his various books, added a great deal to our knowledge of marine [animals], many of them microscopic; and this book is amongst his best on this subject."—*Guardian*, June 11, 1856.

" This charming issue from his fertile pen will delight scores of naturalists, as well as induce a liking for a healthy and rational amusement among the many loungers who indulge in a sea-side holiday."—*Lincolnshire Times*, June 10, 1856.

LONDON: JOHN VAN VOORST, PATERNOSTER ROW.

THE MARINE AQUARIUM.

(SALE LIST.)

19 & 20, PORTLAND ROAD,

REGENT'S PARK,

LONDON.

MR. W. ALFORD LLOYD begs to announce that he has made very extensive arrangements for the sale of Living Marine Animals, Sea-Weeds, Tanks, and all other accessories for the study of Aquarian Natural History.

Mr. LLOYD ordinarily keeps a stock of fifteen thousand specimens, comprising two hundred genera, acclimated in fifty large plate-glass tanks, aggregating more than a thousand gallons of sea water. The peculiarity which distinguishes this collection above that which any other single spot can furnish, and which renders it an object of attention not only to the amateur and student residing in London and in other inland places, but also to naturalists living at distant parts of the coast, is, that it is the result of an organized body of gatherers, posted at intervals in the richest localities; and thus our Marine Fauna and Flora are very adequately represented, as to variety, in the Metropolis, and may be had from thence much more advantageously than from the coast direct, inasmuch as the specimens are selected with the special view of their respective capabilities of enduring confinement. Arrangements with foreign correspondents are also in course of progress. The most delicate organizations may be sent by rail or by post with perfect safety.

The discovery of a mode of making Artificial Sea-Water gives large facilities for the successful prosecution of the study. Much time, therefore, has been spent in assimilating it to the condition of the actual water of the ocean, so that it is offered as an analytically correct compound, which thoroughly answers every purpose, and which improves in good qualities the longer it is kept in use unchanged. Thus the permanent maintenance of a collection of Living Marine Animals and Algæ is rendered a far more easily attainable matter than even the domestic culture of flowers. To render this yet more practicable in the hands of inexperienced persons, Mr. LLOYD makes it a point to keep in stock great numbers of small portable Aquaria ready stocked, and with the balance of existence properly adjusted.

Although from their nature the inhabitants of the ocean have a greater interest than Fresh-Water collections, the latter are duly provided and are stocked with appropriate inhabitants, both vertebrate and invertebrate, many of them with a view of more accurately observing the habits of those creatures which have hitherto been only imperfectly preserved in cabinets, or which, from their perishable nature, cannot be preserved at all, except in a living state. Various arrangements have also been adopted so as to combine the Aquarium with methods of growing Ferns, Mosses, Lichens, &c., and to fit them for the study of the habits, embryology, and development of Semi-Aquatics, both animal and vegetable. A list of this department is in preparation.

The Tanks are constructed exclusively by the eminent firm (for whom Mr. LLOYD is sole Agent) of SANDERS and WOOLCOTT, Makers to the Zoological Societies of London and Ireland; to his Grace the Duke of Devonshire; to the Right Hon. Sir Robert Peel; and to various public and private collections throughout the kingdom. As at present improved, these are not merely vessels for the reception of Plants and Animals, devised without reference to a purpose; for a long series of observations on the scientific requirements demanded of them, has so perfected them, that they very accurately imitate natural conditions by attention being paid to the direction, intensity, and colour of the light employed; by the furnishing of various depths and densities of the water, by the regulation of the temperature, and by the arrangement of the whole for special purposes. Nor has external decoration been neglected. As complete and independent pieces of furniture, many are made of ornamental woods, are mounted table-height, and are placed on castors, for the facility of being easily moved when full to any part of the room or house, as the aspect of the sun or the time of the year may demand.

The wants of Microscopists are met by the preservation, for the use of that class of observers, of such organisms (both living and dead) as are not otherwise easily attainable.

To render Mr. LLOYD's establishment as complete as possible, the literary portion of the subject has received prominent notice, and all the Books and Periodicals in any way allied to it, are recognized as forming a portion of the stock, and are laid upon the tables for the use of visitors, and for purposes of sale.

LIST.

Real sea water. 6d. per gallon.

Marine salts, for the production of artificial sea water,—a pound makes nearly three gallons. 1s. per lb.

Specific gravity test, for regulating density of the water. 1s.

Thermometer for regulating temperature, with the bulb arranged for constant immersion. 3s. 6d.

Pocket magnifiers. } 2s. 6d.
Coddington lens. } to
Stanhope lens. } 12s. 6d.

Gairdner's Microscope.

Warington's compound microscope, arranged for viewing objects in an Aquarium. 103s.

Drip glasses for aeration. 2s. 6d.

Syringe for aeration. 1s. 6d.

Dipping tubes and spoons for the removal of offensive matters. 6d. to 1s.

Gutta percha and other siphons for drawing off water without disturbance. 3s. 6d.

Stone travelling one-gallon jars with top, packed in wicker. 3s. 6d.

SEA WEEDS.

GREEN.
4d. to 8d.

Ulva latissima.
,, lactuca.
Entermorpha intestinalis
,, compressa.
Cladophora arcta.
,, rupestris.
Bryopsis plumosa.

RED.
6d. to 1s.

Iridæa edulis.
Griffithsia setacea.
Delesseria sanguinea.
,, alata.
Corallina officinalis.
Rhodomela subfusca.
Gracilaria confervoides.
Gelidium corneum.
Chondrus crispus.
Phyllophora rubens.
Polyides rotundus.
Ceramium rubrum.

ZOOPHYTES.

MADREPORES.
1s. to 2s.

Caryophyllia Smithii.
Balanophyllia regia.

SEA-ANEMONES.
1s. to 3s. 6d.

Sagartia viduata=angui-
coma.
,, troglodytes (five
varieties).
,, aurora.
,, candida.
,, miniata.

1s. to 7s.

Sagartia rosea.
,, nivea.
,, venusta.
,, parasitica.
,, bellis (four va-
rieties.
,, dianthus (five
varieties).

5s.

Sagartia aurantiaca.
,, pulcherrima.

1s. to 2s. 6d.

Bunodes alba.
,, gemmacea.
,, thallia.
,, clavata.

ZOOPHYTES.
SEA-ANEMONES.
Continued.

6d. to 1s.

Bunodes crassicornis
(three varieties.)

6d. to 1s. 6d.

Actinia mesenbryanthe-
mum (four varieties).

1s. to 7s.

Anthea cereus (three va-
rieties).
Adamsia palliata.
Edwardsia sphæroides.
,, vestita.
Corynactis viridis.

NAKED AND TUBED HYDROIDA.
6d. to 1s.

Clava multicornis.
Hydractinia echinata.
Coryne pusilla.
Tubularia indivisa.
Sertularia polyzonias.
,, abietina.
,, filicula.
,, cupressina.
Thuiaria thuia.
Antennularia antennina.
Campanularia volubilis.
Laomedea geniculata.

STAR-FISHES AND SEA URCHINS.
6d. to 1s.

Uraster rubens.
Asterina gibbosa.
Goniaster equestris.
Echinus miliaris.
,, sphæra.

SEA CUCUMBERS.
6d. to 2s.

Pentactes pentacta.
Ocnus brunneus.

TUBE AND OTHER WORMS.
6d, to 2s. 6d.

Sabellaria alveolata.
Sabella ventilabrum.
,, reniformis.
,, tubularia.
Serpula contortuplicata
(four varieties).
Serpula triquetra (three
varieties).
Terebella conchilega.

TUBE AND OTHER WORMS.—*Continued.*

6d. to 2s. 6d.

Pectinaria Belgica.
Spirorbis communis.
Spio seticornis.
Pontobdella muricata.
Aphrodita aculeata.
Nereis bilineata.
,, pelagica.
Phyllodoce viridis.

CRUSTACEA.
6d. to 1s. 6d.

Idotæa appendiculata.
Palæmon serratus.
,, Leachii.
,, squilla.
Crangon vulgaris.
Hippolyte Thompsoni.
Porcellana platycheles.
Pagurus Bernhardus.
,, Prideauxii.
Carcinus Mænas.
Cancer pagurus.
Portunus depurator.
Xantho florida.

BARNACLES.
6d. to 1s.

Balanus balanoides.
Pyrgoma Anglicum (on
Madrepores.)

POLYZOA.
6d. to 1s.

Bowerbankia imbricata.
Vesicularia spinosa.
Serialia lendigera.
Lepralia (several species)
Membranipora pilosa.

MOLLUSKS.
6d. per doz. to 1s. 6d. each.

Nassa reticulata.
Murex erinaceus.
Litorina litorea.
,, rudis.
Natica monilifera.
Purpura lapillus.
Rissoa (several species)
Trochus cinereus.
,, ziziphinus.
Haliotis tuberculata.
Fissurella reticulata.
Patella vulgata.
Dentalium entalis.
Ostrea edulis.
Anomia ephippium.

MOLLUSKS.	MOLLUSKS.	FISHES—continued.
Continued.	*Continued.*	6d. to 2s.
6d. per doz. to 1s. 6d. each.	6d. per doz. to 1s. 6d. each,	Cottus quadricornis.
Doris	Pholas dactylus.	Gobius niger.
Ægirus ⎫	Ascidia virginea, &c.	,, unipunctatus.
Ancula ⎬ and other	Cynthia quadrangularis,	,, minutus.
Tritonia ⎭ Nudibranchs	&c.	,, Ruthensparri.
Eolis ⎭ in season.	Botryllus polycyclus, &c.	Syngnathus acus.
Aplysia hybrida.		,, typhle.
Pecten maximus.		,, lumbriciformis
,, opercularis.	**FISHES.**	Murænoides guttata.
,, varius.	6d. to 2s.	Blennius ocellaris.
Mytilus edulis.		,, pholis.
Modiola modiolus.	Gasterosteus spinachia.	Labrus maculatus.
,, barbata.	Cottus scorpius.	Crenilabrus cornubicus.
Saxicava rugosa.	,, bubalis.	

Nearly all of the foregoing marine list are kept on hand, but there are some which can be procured to order.

N.B.—In order to ensure success on the part of inexperienced amateurs, Mr. LLOYD reserves to himself a discretionary power of selecting stock, and in supplying them of such kind, and in such number and order of sequence as may seem to him most desirable.

TANKS, &c.

Size	ft. in.	ft. in.	ft. in.		£ s. d.
Size . .	6 0 long,	2 0 wide,	2 6 deep	all glass . .	21 0 0
,, . .	6 0 ,,	2 0 ,,	2 6 ,,	solid ends .	18 0 0
,, . .	3 0 ,,	1 4 ,,	1 6 ,,	ditto .	5 0 0
,, . .	3 0 ,,	1 4 ,,	1 6 ,,	all glass . .	5 5 0
,, . .	2 4 ,,	1 4 ,,	1 3 ,,	ditto .	3 15 0
,, . .	2 2 ,,	1 0 ,,	1 3 ,,	ditto .	3 3 0
,, . .	1 6 ,,	0 10 ,,	1 0 ,,	ditto .	1 11 6
,, . .	1 4 ,,	0 9½ ,,	0 9 ,,	ditto .	1 0 0

Octagon Aquarium, according to size, varying from 3 0 0
Aquarium, with Fern Case,   ditto   ditto . . . 3 10 0
,,   adapted for Leeches, ditto   ditto . . 1 10 0
,,   on Warington's Slope-Back principle 3 0 0
Tanks fitted on legs (table height) with castors, as independent
    articles of furniture, *en suite*, according to style required . .
Shallow Rock Pools, as suggested by Mr. Warington . . . 1 10 0
Vase Aquarium, arranged to turn round on its stand . 5 0 to 12 0
Cylindrical Glasses, with or without coloured glass for the⎫
    growth of Rhodosperms, as proposed by Mr. Warington ⎬ 1 6 to 3 0
Shallow Glass Pans . . . . ⎭
Smith and Beck's Zoophyte Tanks . . . . . 10 6 to 16 0
,,   ,,   Chara Troughs 7 6

*** DELIVERY AND PACKING CASES ACCORDING TO CIRCUMSTANCES.

Money Orders to be made payable at 103, Tottenham Court Road. Amounts under 10s. may be remitted in stamps.

Crystal Palace Guides series

A facsimile series published by Euston Grove Press

www.EustonGrove.com

Lightning Source UK Ltd.
Milton Keynes UK
02 March 2011

168506UK00001B/93/P